ANIMAL GENIUSES

BONOBOS

THERESA EMMINIZER

PowerKiDS press

Published in 2026 by The Rosen Publishing Group, Inc.
2544 Clinton Street, Buffalo, NY 14224

First Edition

Editor: Theresa Emminizer
Book Design: Michael Flynn

Photo Credits: Cover, pp. 1, 5, 9, 13 GUDKOV ANDREY/Shutterstock.com; (series background) Viktoriia LiSa/Shutterstock.com; pp. 7, 15 (bonobo), 17, 21 Sergey Uryadnikov/Shutterstock.com; p. 7 (orangutan) Al Carrera/Shutterstock.com; p. 7 (chimpanzee) Kjersti Joergensen/Shutterstock.com; p. 7 (gorilla) svetjekolem/Shutterstock.com; p. 11 Edwin Butter/Shutterstock.com; p. 15 Martin Harvey/Avalon.red/ Alamy Stock Photo; p. 19 https://commons.wikimedia.org/wiki/File:Kanzi,_conversing.jpg.

Cataloging-in-Publication Data

Names: Emminizer, Theresa.
Title: Bonobos / Theresa Emminizer.
Description: Buffalo, NY : PowerKids Press, 2026. | Series: Animal geniuses | Includes glossary and index.
Identifiers: ISBN 9781499451764 (pbk.) | ISBN 9781499451771 (library bound) | ISBN 9781499451788 (ebook)
Subjects: LCSH: Bonobo--Juvenile literature. | Bonobo--Behavior--Juvenile literature. | Animal intelligence--Juvenile literature.
Classification: LCC QL737.P94 E45 2026 | DDC 599.885--dc23

Manufactured in China

CPSIA Compliance Information: Batch #QSPK26. For Further Information contact Rosen Publishing at 1-800-237-9932.

CONTENTS

Meet the Bonobo!

Say "hello" to the bonobo! Looking at a bonobo's furry face, you might be surprised to hear that you share a lot in common with this animal. In fact, bonobos are some of our closest animal relatives, or family members! They share nearly all of our **DNA**. They're also supersmart!

Great Apes

Bonobos are great apes. Apes are tailless **primates**. Chimpanzees, gorillas, and orangutans are great apes too! Bonobos live in the African **rainforest** in a country called the Democratic Republic of Congo. They aren't native to any other parts of Africa or the world!

bonobo

orangutan

chimpanzee

gorilla

Bonobo or Chimp?

People used to think that bonobos were a kind of chimp. That's not true! Bonobos and chimps look alike, but bonobos have smaller heads. They walk up on two feet more often than chimps. The hair on bonobos' heads is parted down the middle. Bonobos also behave, or act, very differently from chimps.

Peaceful Partners

Bonobos are peaceful animals. This makes them very different from chimpanzees and humans, who often fight and harm one another. No bonobo has ever been known to kill another. Instead, bonobos find other ways to **solve** problems. Bonobos don't fight over territories, or places to live, the way chimpanzees do.

Female Leaders

Bonobos live in groups called troops. Bonobo troops are usually led by females, or girls. If a male bonobo acts aggressive, or tries to fight, the females work together to stop the male from harming anyone. Female bonobos help each other birth their babies!

Working Together

Bonobos use teamwork to gather food. They share their food with the troop. Bonobos also use tools! They use twigs, or small branches, to dig up bugs or clean their teeth. They use moss to gather water from trees. Adult bonobos teach these things to their young.

Building Together

Bonobos help each other build nests in treetops using leaves and twigs. Sometimes adult bonobos share nests together. That's unusual for great apes! Bonobos sleep in big groups so they can help each other look out for danger. They call out to each other if they feel unsafe.

Smart Talk

Scientist Sue Savage-Rumbaugh taught a bonobo named Kanzi to understand words. Kanzi can **communicate** and answer questions by pointing to **symbols** on a board. Kanzi knows 348 symbols and even more words. Kanzi can also use matches to start a fire and toast marshmallows on a stick!

Beautiful Bonobos

Bonobos are a lot like people! They play and laugh together. They even tickle each other! Scientists say that bonobos are some of the smartest animals on Earth. There's much we still don't know about them! What more could we learn from our amazing primate relatives?

GLOSSARY

communicate: To share ideas and feelings through sounds and motions.

DNA: A matter that carries genetic information in a plant or animal's cells.

primate: The group of animals that includes monkeys, gorillas, and humans.

rainforest: A tropical woodland with heavy rainfall and marked by leafy trees.

scientist: Someone who studies or practices science.

solve: To find an answer to a problem.

symbol: Something that stands for something else.

FOR MORE INFORMATION

BOOKS

Eason, Sarah. *Kanzi Learns Language! Supersmart Ape*. Minneapolis, MN: Bearport Publishing, 2023.

Hudak, Heather C. *Gorilla, Baboon, and Other Troops*. St. Catharines, Ontario: Crabtree Publishing, 2023.

WEBSITES

Britannica Kids: Bonobo
kids.britannica.com/kids/article/bonobo/433076
Find out more about what makes bonobos so special!

Friends of Bonobos
www.bonobos.org/bonobos
Learn more about bonobos.

Publisher's note to educators and parents: Our editors have carefully reviewed these websites to ensure that they are suitable for students. Many websites change frequently, however, and we cannot guarantee that a site's future contents will continue to meet our high standards of quality and educational value. Be advised that students should be closely supervised whenever they access the internet.

INDEX